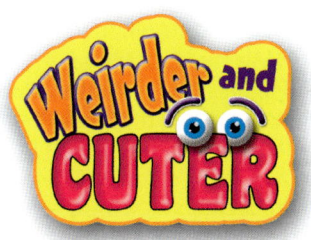

Axolotl

by Dawn Bluemel Oldfield

Consultant: Darin Collins, DVM
Director, Animal Health Programs
Woodland Park Zoo
Seattle, Washington

New York, New York

Credits

Cover, © Jane Burton/Minden Pictures and © DeshaCAM/Shutterstock; TOC, © Eric Isselee/Shutterstock; 4–5, © Juniors Bildarchiv GmbH/Alamy Stock Photo; 6, © Tremor Photography/Shutterstock; 7, © Bruno Cavignaux/Minden Pictures; 8T, © Manfred Ruckszio/Shutterstock; 8B, © Lapis2380/Shutterstock; 9, © blickwinkel/Hartl/Alamy Stock Photo; 10T, © Pete Oxford/Minden Pictures; 10B, © blickwinkel/Hartl/Alamy Stock Photo; 11,© Roberto Nistri/Alamy Stock Photo; 12L, © Evgeny Karandaev/Shutterstock; 12–13, © AndreyTTL/iStock; 13T, © Argument/iStock; 13B, © aureapterus/iStock; 14L, © Pan Xunbin/Shutterstock; 14R, © furtseff/Shutterstock; 15, © Andrea Izzotti/Shutterstock; 16, © Rostislav Stefanek/Shutterstock; 17, © Tatiana Gordievskaia/Shutterstock; 18T, © Lapis2380/Shutterstock; 18B (L to R), © David Gardiner; 19, © TOMAS BRAVO/REUTERS/Alamy Stock Photo; 20, © Christian Hütter/imageBROKER/Alamy Stock Photo; 21, © Miroslava Kopecka/Dreamstime; 22 (T to B), © Europics/Newscom, © Bill Peterman, and © Daniel Heuclin/Nature Picture Library/Alamy Stock Photo; 23TL, © Alen thien/Shutterstock; 23TR, © Argument/iStock; 23BL, © D. Kucharski K. Kucharska/Shutterstock; 23BR, © Tremor Photography/Shutterstock.

Publisher: Kenn Goin
Editor: Jessica Rudolph
Creative Director: Spencer Brinker
Design: Debrah Kaiser
Photo Researcher: Thomas Persano

Library of Congress Cataloging-in-Publication Data

Names: Bluemel Oldfield, Dawn, author.
Title: Axolotl / by Dawn Bluemel Oldfield.
Description: New York, New York : Bearport Publishing, [2018] | Series:
  Weirder and cuter | Audience: Ages 5–8.
Identifiers: LCCN 2017012313 (print) | LCCN 2017019437 (ebook) | ISBN
  9781684023158 (ebook) | ISBN 9781684022618 (library)
Subjects: LCSH: Axolotls—Juvenile literature. | Amphibians—Juvenile
  literature.
Classification: LCC QL668.C23 (ebook) | LCC QL668.C23 B58 2018 (print) | DDC
  597.8/58—dc23
LC record available at https://lccn.loc.gov/2017012313

Copyright © 2018 Bearport Publishing Company, Inc. All rights reserved. No part of this publication may be reproduced in whole or in part, stored in any retrieval system, or transmitted in any form or by any means, electronic, mechanical, photocopying, recording, or otherwise, without written permission from the publisher.

For more information, write to Bearport Publishing Company, Inc., 45 West 21st Street, Suite 3B, New York, New York 10010. Printed in the United States of America.

10 9 8 7 6 5 4 3 2 1

Contents

Axolotl 4

More Weird Salamanders 22

Glossary..................... 23

Index 24

Read More 24

Learn More Online 24

About the Author 24

What's this weird but cute **amphibian**?

It's an **axolotl** (ak-suh-LAH-tuhl).

Tiny **eye**s!

The axolotl is a type of amphibian called a **salamander**.

Many salamanders spend time on land.

Yet axolotls spend their whole lives underwater!

They live in a lake in Mexico.

fire salamander

The axolotl has two nicknames—the Mexican salamander and the walking fish.

Axolotls look like tadpoles with legs.

tadpole

fin

The axolotl's fin and strong tail help it swim.

These odd animals have big, feathery **gills**!

They use their gills to breathe underwater.

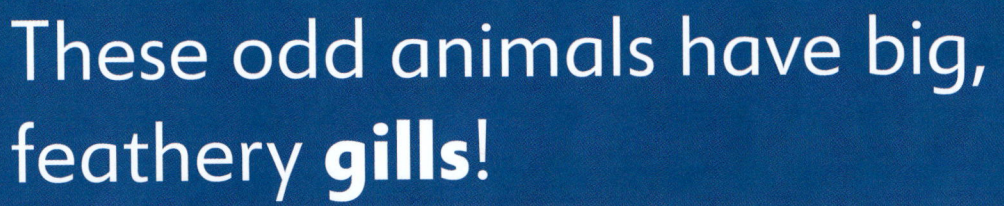

gills

Baby axolotls grow inside tiny eggs.

After two to three weeks, the babies hatch.

They grow front legs first, and then back legs.

The mother axolotl lays her eggs on leaves or rocks in the water. She lays more than 1,000 eggs!

front leg

Axolotls grow to be 8 to 12 inches (20 to 30 cm) long.

They weigh about 2 to 8 ounces (57 to 227 g).

That's less than a can of soda.

Axolotls can have yellow, white, black, or speckled skin.

Axolotls dart in the water to hunt for **prey**.

They eat worms, small fish, and insects.

The salamanders catch food with tiny stump-like teeth.

Look out!

Large fish hunt axolotls.

Axolotls can be harmed by pollution, too. The salamanders need to live in water that is very clean.

Birds, such as herons, also eat the small creatures.

Did you know axolotls have a special healing ability?

If an axolotl loses a leg to an enemy, it can grow a new one!

Growing a New Leg

an axolotl scientist in Mexico

Scientists are studying these amazing amphibians to learn more about how they regrow body parts.

Can you believe some people have axolotls as pets?

They keep the little animals in fish tanks filled with water.

The owners feed them worms and shrimp.

More Weird Salamanders

Chinese Giant Salamander

The Chinese giant salamander lives in streams and lakes in China. It's the largest amphibian in the world. It can grow to be 5 feet 9 inches (1.8 m) long!

Patch-Nosed Salamander

This is the world's smallest salamander. It was discovered in Georgia in 2009. It's about the size of a dime.

Texas Blind Salamander
The Texas blind salamander can grow to be 5 inches (13 cm) long. It lives in dark caves in San Marcos, Texas, and it has no eyes!

Glossary

amphibian (am-FIB-ee-uhn) an animal that lives in water while it's young; most live on land as adults

gills (GILZ) the body parts of underwater animals that are used for breathing

prey (PRAY) an animal that is hunted by another animal for food

salamander (SAL-uh-*man*-dur) a type of amphibian with smooth, moist skin that looks like a lizard

Index

colors 13
eggs 10–11
enemies 16–17, 18
gills 5, 9
Mexico 6–7, 19
prey 14–15
scientists 19
size 12, 22
swimming 8

Read More

Goldish, Meish. *Slimy Salamanders (Amphibiana).* New York: Bearport (2010).

Mason, Susan. *Axolotl!: Fun Facts about the World's Coolest Salamander.* London: Bubble Publishing (2016).

Learn More Online

To learn more about axolotls, visit
www.bearportpublishing.com/WeirderandCuter

About the Author

Dawn Bluemel Oldfield is a writer who enjoys reading, traveling, and gardening. She and her husband live in Texas, where, sadly, there are no axolotls.